Science

Practice for the
West Virginia *WESTEST*

Grade 5

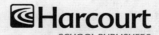

Harcourt
SCHOOL PUBLISHERS

Orlando   Austin   New York   San Diego   Toronto   London

Visit *The Learning Site!*
www.harcourtschool.com

Contents

This page and the next give examples of the two types of items found on the science portion of *WESTEST*.

Multiple-Choice Items

Most of the items on the science portion of *WESTEST* are multiple-choice. Each multiple-choice item has four answer choices. The tips that follow will help you answer these questions.

1. Read the question carefully. Restate the question in your own words.

2. Watch for key words such as **best, most,** or **least.**

3. The question might include one or more tables, graphs, diagrams, or pictures. Study these carefully before choosing an answer.

4. Find the best answer for the question. Fill in the answer bubble for that answer. Do not make any stray marks around answer spaces.

1. This diagram shows part of an aquatic food web.

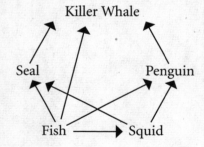

What is missing from this food web?

(A) producer

(B) first-level consumer

(C) second-level consumer

(D) top-level consumer

2-Point Constructed-Response Items

For some items, you must write a brief answer to explain a science concept or to apply a science process skill. If a portion of your answer is correct, you will earn 1 point. If your answer is complete and correct, you will earn a score of 2 points.

1. Martin is making a model of energy levels for a typical forest ecosystem. He will draw pictures of producers, herbivores, and carnivores to show the kinds of organisms that exist at each of the levels.

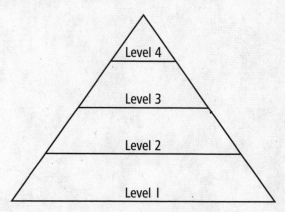

In which energy level or levels should he draw pictures of producers? Explain.

_____

_____

_____

In which energy level or levels should he draw pictures of carnivores? Explain.

_____

_____

_____

Practice Sets

Each practice set contains five items. Some items have Tips, which give clues about how to answer the items. An item may have a graph, table, picture, or diagram. Study these carefully before answering the items.

Name _____

Date _____

1. Which tool could you use to take measurements
    in units called newtons?

Ⓐ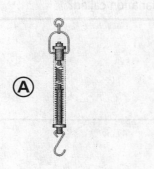

Ⓒ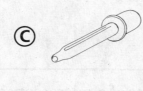

Ⓑ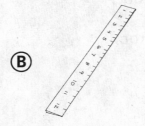

Ⓓ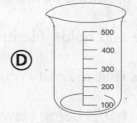

2. You are working in a science lab, and you accidentally
    break a glass jar. What should you do first?

    Ⓐ Ask your partner to help you.

    Ⓑ Sweep the glass into a trash can.

    Ⓒ Tell your teacher about the break.

    Ⓓ Wash your hands with soap and water.

3. What is tested during an experiment?

(A) data

(B) conclusion

(C) hypothesis

(D) variable

> **Tip**
>
> Before you start an experiment, you should suggest an outcome or explanation that can be tested. What is this outcome or explanation called?

---

4. What is the purpose of the *stage* of a microscope?

(A) It contains one lens and is mounted at the end of a tube.

(B) It supports the microscope and holds a lamp or mirror.

(C) It holds one or more lenses that magnify objects.

(D) It holds the slide or object that is being viewed.

© Harcourt

5. Two common temperature scales are the Celsius scale and the Fahrenheit scale. Look at the two temperature scales on the thermometer below.

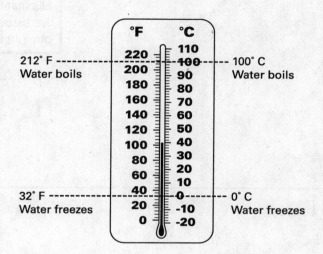

212° F --------- Water boils --------- 100° C Water boils

32° F --------- Water freezes --------- 0° C Water freezes

On the Celsius scale, the freezing point of water is 0°C, and the boiling point is 100°C. What are the freezing and boiling points of water on the Fahrenheit scale?

_____

_____

Suppose you have two beakers of water at the same temperature, and you warm them both. You raise the temperature of one by 5 degrees Celsius, and you raise the temperature of the other by 5 degrees Fahrenheit. Which beaker of water is warmer?

_____

_____

_____

_____

Name _____

Date _____

1. What type of tissue makes up most of your skin?

   Ⓐ connective

   Ⓑ epithelial

   Ⓒ muscle

   Ⓓ nervous

> **Tip**
> Eliminate terms that are tissues with other functions.

2. What attaches muscles to bones?

   Ⓐ capillaries

   Ⓑ cartilage

   Ⓒ ligaments

   Ⓓ tendons

3. Which cell part contains information about
the characteristics of the cell?

Ⓐ cell wall

Ⓑ chloroplast

Ⓒ chromosome

Ⓓ cytoplasm

> **Tip**
> Narrow your choices
> by looking for parts
> that are common to
> all cells.

4. Which organ system is made up of the heart,
blood vessels, and the blood?

Ⓐ circulatory system

Ⓑ digestive system

Ⓒ excretory system

Ⓓ nervous system

Name _____ Date _____

5. When you inhale, oxygen in the air you breathe enters your respiratory system. Look at the respiratory system in the illustration.

Tip
Use the diagram to help you answer the question.

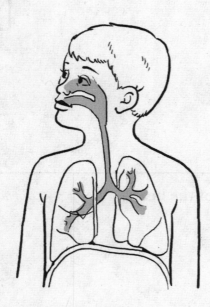

What is the function of the respiratory system?

_____

_____

Where are the alveoli, and what is their function in this system?

_____

_____

_____

_____

_____

© Harcourt

Name _____

Date _____

1. In which of the following classification groups
    are organisms **most closely** related?

    Ⓐ class

    Ⓑ family

    Ⓒ genus

    Ⓓ order

2. A beetle and a spider can both be classified in which
    of the following groups?

    Tip
    Use the pictures of the
    beetle and spider to
    help you classify them
    according to their
    characteristics.

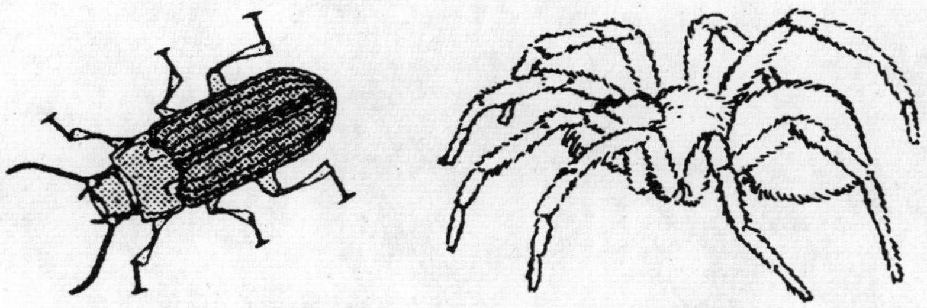

    Ⓐ arthropods

    Ⓑ crustaceans

    Ⓒ echinoderms

    Ⓓ insects

© Harcourt

Name _____ Date _____

3. Which animal is a vertebrate?

    Ⓐ butterfly

    Ⓑ lobster

    Ⓒ snake

    Ⓓ sponge

> **Tip**
> What does the term
> *vertebrate* mean?

4. Which group of organisms is sometimes separated into
two different kingdoms?

    Ⓐ bacteria

    Ⓑ fungi

    Ⓒ plants

    Ⓓ protists

© Harcourt

5. When classifying organisms, scientists consider both structure and function. Look at the illustration below of a grasshopper.

What is the difference in structure and function?

_____

_____

Name a function of the grasshopper's back legs.

_____

© Harcourt

1. Which term describes the type of root shown in this drawing?

Tip
This root firmly anchors the plant in the ground.

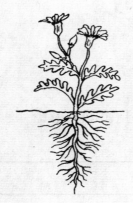

(A) runner

(B) taproot

(C) prop root

(D) fibrous root

2. During a fern's life cycle, male and female reproductive cells join to form what type of cell?

(A) gamete

(B) sperm

(C) spore

(D) zygote

© Harcourt

3. While taking a walk through the park, you see a tree with colorful blossoms. How would you classify this tree?

Tip
A tree with blossoms is a type of flowering plant. Which answer choice refers to flowering plants?

Ⓐ angiosperm

Ⓑ gametophyte

Ⓒ gymnosperm

Ⓓ sporophyte

4. Through which part does the leaf take in carbon dioxide?

Ⓐ epidermis

Ⓑ phloem

Ⓒ stomata

Ⓓ xylem

© Harcourt

5. Plants can be described as vascular or nonvascular. Look at the plants shown below.

fern

moss

rose

spruce tree

Explain what the term *vascular* means.

_____

_____

Classify each of the plants shown as vascular or nonvascular.

_____

_____

© Harcourt

Name _____

Date _____

1. A certain kind of frog has 20 chromosomes in each body cell. How many chromosomes does the frog have in its gametes?

Tip
What is the function of *gametes*?

Ⓐ 5

Ⓑ 10

Ⓒ 20

Ⓓ 40

2. The illustration shows a stage of mitosis.

Which term refers to the system of thin tubes that form across the cell?

Ⓐ centriole

Ⓑ membrane

Ⓒ nucleus

Ⓓ spindle

© Harcourt

3. Which is the **best** description of how body cells help an organism grow?

   Ⓐ They become larger.

   Ⓑ They undergo meiosis.

   Ⓒ They divide to make more cells.

   Ⓓ They form cells with twice as many chromosomes.

> **Tip**
> Look for the result that makes sense with what you already know and have observed.

4. Anita has naturally red hair. Her parents and brother do not have red hair. Which term describes the genetic cause of red hair?

   Ⓐ dominant trait

   Ⓑ diseased gene

   Ⓒ recessive trait

   Ⓓ environmental factor

© Harcourt

5. Study the diagram below, which shows information
about the blue mussel.

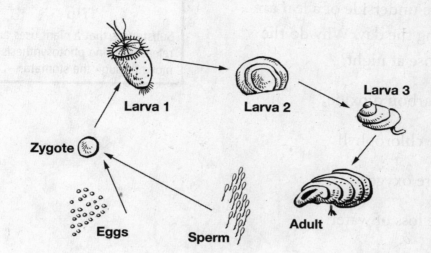

Write an appropriate title for this diagram.

_____

_____

The blue mussel looks different as it changes from
a larva into an adult. What is this change in
appearance called?

_____

Name _____

Date _____

1. The stomata on the underside of a leaf are usually open during the day. Why do the stomata usually close at night?

Tip
Substances that a plant uses and releases during photosynthesis move through the stomata.

(A) to get rid of carbon dioxide

(B) to make more chlorophyll

(C) to take in more oxygen

(D) to prevent the loss of water

2. Every time you eat a meal, you gain energy. What is the original source of this energy?

(A) the sun

(B) the food web

(C) consumers

(D) decomposers

© Harcourt

3. Which function is carried out by plants and affects the air you breathe?

Ⓐ Plants help remove smoke from the air.

Ⓑ Along with the wind, plants help mix the air.

Ⓒ Plants exchange carbon dioxide for oxygen in the air.

Ⓓ Plants release nitrogen during photosynthesis.

4.

For a school experiment, Stella places a plastic bag over one leaf of a houseplant. At the end of the day, the inside of the bag is covered in water droplets. Which process produced the droplets?

Ⓐ absorption

Ⓑ photosynthesis

Ⓒ respiration

Ⓓ transpiration

© Harcourt

5. The food web below shows the relationships among some organisms that live in a wooded area.

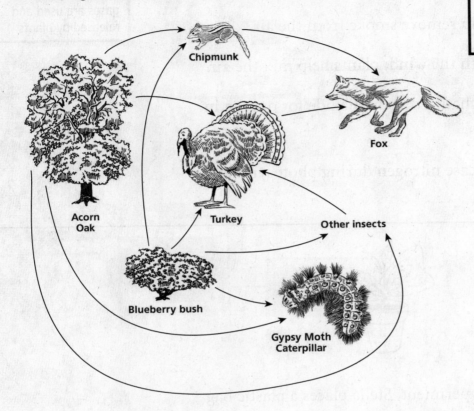

Categorize each organism as a producer, a herbivore, an omnivore, or a carnivore.

_____

_____

_____

Describe how the fox depends on the gypsy moth caterpillar.

_____

_____

_____

_____

© Harcourt

1. The graph below shows major causes of wetland loss in the United States from 1986 to 1997.

Tip
Read labels on the graph and compare the causes of wetland loss.

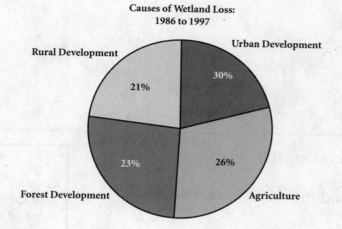

Causes of Wetland Loss:
1986 to 1997

Rural Development

Urban Development

21%

30%

23%

26%

Forest Development

Agriculture

According to the graph, which of the following is a true statement about causes of wetland loss?

(A) More was lost to rural development than to agricultural development.

(B) More was lost to forest development than to rural development.

(C) More was lost to rural development than to urban development.

(D) More was lost to forest development than to urban development.

© Harcourt

2. Which term describes all of the organisms in an ecosystem?

    Ⓐ community

    Ⓑ competition

    Ⓒ population

    Ⓓ succession

3. Which of these species is at greatest risk of extinction?

> **Tip**
> Consider how easily each species could either adapt or move to another area.

    Ⓐ A plant species found in several areas of a few different states.

    Ⓑ A small animal species found in a forest that was partly destroyed by fire.

    Ⓒ A plant species found only on an island infected by a disease harmful to the plant.

    Ⓓ A small animal species located only in an area devastated by a volcano.

4. Fish use their gills to get dissolved oxygen from water. Aquarium owners must be sure that the number of fish can be supported by the amount of dissolved oxygen in an aquarium. In this situation, what is dissolved oxygen an example of?

(A) adaptation

(B) competition

(C) limiting factor

(D) succession

5. *Remoras* are sucker-fish. The dorsal (back) fin of the remora forms a sucker. This allows a remora to attach itself to a shark. The small remora "hitches a ride" and feeds on scraps of food left by the shark. The shark is unaffected. The relationship between a remora and a shark is an example of which type of symbiosis? How do you know?

> **Tip**
> Think about how organisms help or harm each other by symbiosis.

_____

_____

_____

Name two other types of symbiosis, and give an example of each.

_____

_____

© Harcourt

Name _____

Date _____

1. Sylvia has found a rock that has seashells in it.
    What type of rock has Sylvia **most likely** found?

    Tip
    Think about which
    type of rock forms
    where shellfish live.

    (A) igneous

    (B) metamorphic

    (C) sedimentary

    (D) volcanic

2. A class uses the following table to identify a set of
    unlabeled minerals.

Mineral	Color	Hardness	Luster	Streak
Apatite	yellow	5	glassy	white
Galena	silver	2.5	metallic	gray
Hematite	silver	5	metallic	dark red
Magnetite	black	6	metallic	black
Pyrite	yellow	6	metallic	dark green

    What property will be **most** useful for identifying all of
    the minerals?

    (A) color

    (B) hardness

    (C) luster

    (D) streak

© Harcourt

3. Lin found an igneous rock with large mineral crystals. Where did this rock **most likely** form?

Ⓐ in a glacier

Ⓑ deep underground

Ⓒ on the bottom of a river

Ⓓ on the surface of Earth

Tip
Large crystals form when melted rock cools slowly. Where does this happen?

4. The mineral fluorite has a classification of 4 on the Mohs' hardness scale.

Object	Hardness
Fingernail	2.5
Penny	3
Steel nail	5.5
Glass	6

Based on information in the table, what would you probably observe if you tested fluorite's hardness?

Ⓐ Fluorite can scratch a steel nail, but it can be scratched by glass.

Ⓑ Fluorite can scratch a penny, but it can be scratched by glass.

Ⓒ Fluorite can scratch both a steel nail and glass.

Ⓓ Fluorite can be scratched by both a fingernail and a penny.

5. Bryce models different stages of the rock cycle.

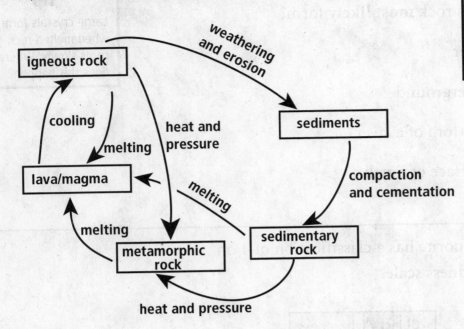

Bryce rubs chalk against a piece of sandpaper. What part of the rock cycle does he model? What stage of the rock cycle would most likely follow next?

_____

_____

_____

For another part of the model, Bryce lets melted butter cool in a bowl. What part of the rock cycle does this step represent?

_____

_____

© Harcourt

Name _____

Date _____

1.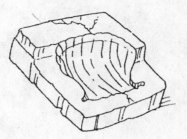

Tip
Focus on the key words "leaving a hollow space."

The illustration shows a fossil that formed when an organism decayed, leaving a hollow space. Which term describes this type of fossil?

(A) carbon film

(B) cast

(C) mold

(D) petrified wood

2. Similar fossils have been found on both sides of the Atlantic Ocean, even though the organisms could not possibly have crossed an ocean. Which of the following explains this?

(A) carbon film dating

(B) continental drift

(C) index fossils

(D) relative age

© Harcourt

3. Fossils show that the first plants and animals lived in what type of environment?

Ⓐ desert

Ⓑ forest

Ⓒ ocean

Ⓓ swamp

4.

The illustration shows a fossil consisting of amber with an insect inside. What was amber before it hardened and trapped the insect?

Ⓐ dissolved minerals

Ⓑ liquid carbon

Ⓒ tree sap

Ⓓ volcanic rock

© Harcourt

5. Look at the rock cliff in the illustration. Scientists have been able to determine the age of all but the lowest rock layer. The lowest level is different because the land has shifted and the layers are at an angle.

Tip
Think about what fossils tell about the rock in which they are found.

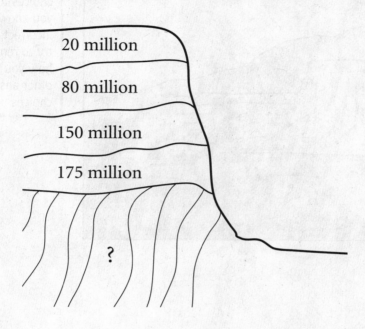

20 million

80 million

150 million

175 million

?

What is an index fossil?

_____

_____

How would an index fossil help scientists date the lowest rock layer?

_____

_____

_____

Name _____

Date _____

1. What is the flat-topped landform shown below?

Tip
Eliminate answer choices that you know are incorrect, then try to remember how you've heard other answer choices used.

Ⓐ mesa

Ⓑ plain

Ⓒ ridge

Ⓓ spit

© Harcourt

2. The table below gives information about some major earthquakes in the world.

Tip
How does an increase of 32 in the strength of an earthquake affect the Richter scale measurement?

Richter Scale Magnitude	Year	Location
8.2	1976	China
8.3	2003	Japan
8.5	1963	Indonesia
9.2	1964	Alaska
9.5	1960	Chile

According to the table, which earthquake was about 32 times the strength of the 1976 earthquake in China?

(A) Japan in 2003

(B) Indonesia in 1963

(C) Alaska in 1964

(D) Chile in 1960

3. Which of the following cause earthquakes?

(A) faults and sinkholes

(B) faults and plate collisions

(C) deposition and plates pulling apart

(D) plates scraping together and erosion

4. Which of the following are forces that create landforms?

   Ⓐ ice and sunlight

   Ⓑ rain and temperature

   Ⓒ water and wind

   Ⓓ wind and heat

5. Scientists in Hawai'i study volcanoes. The volcanoes have broad, gentle slopes, like the volcano shown in the picture.

Tip
Study the picture, and look for key words to help you identify the type of volcano.

What kind of volcanoes are the scientists **most likely** studying?

_____

Name and describe two other types of volcanoes.

_____

_____

_____

_____

© Harcourt

1. Which of the following is a good way to conserve resources?

   Ⓐ Drinking juice from a juice box.

   Ⓑ Drying clothes in an electric dryer.

   Ⓒ Eating dinner on disposable plates.

   Ⓓ Giving outgrown clothes to a resale shop.

2. One method of soil conservation is intercropping, in which farmers plant different crops near each other. Intercropping keeps soil healthy by reducing the need for

   Tip
   Recall that many insects eat only one kind of crop.

   Ⓐ fertilizer.

   Ⓑ pesticides.

   Ⓒ plowing.

   Ⓓ seeding.

© Harcourt

3. What causes most air pollution?

   Ⓐ smoke produced by forest fires

   Ⓑ substances released by burning coal and oil

   Ⓒ pesticides and toxins from household products

   Ⓓ chemicals released by industrial processes

4. Which is a nonrenewable resource?

   Ⓐ water

   Ⓑ trees

   Ⓒ soil

   Ⓓ air

> **Tip**
> Which resource takes thousands of years to form?

Name _____ Date _____

5. Which type of soil conservation is shown in the illustration below?

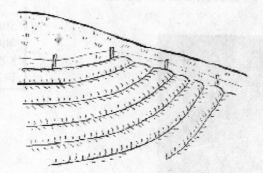

What is the name of this type of soil conservation?

_____

Describe why this type of soil conservation is helpful.

_____

_____

_____

© Harcourt

1. The illustration shows layers of Earth's atmosphere.

Tip
Study the illustration, and apply what you know about weather to find the answer.

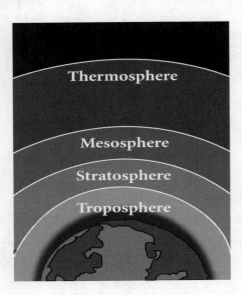

In which layer of the atmosphere does **most** of Earth's weather occur?

Ⓐ thermosphere

Ⓑ mesophere

Ⓒ stratosphere

Ⓓ troposphere

2. Sometimes when you go outside early in the morning, you notice tiny drops of dew on the grass. What causes dew to form?

(A) evaporation of water into the air

(B) condensation of water vapor in the air

(C) condensation of water vapor in the ground

(D) evaporation of water from beneath the ground

3. Which kind of air mass forms over land and brings cool, dry weather?

(A) continental polar

(B) continental tropical

(C) maritime polar

(D) maritime tropical

> **Tip**
> The key words are *land* and *cool*. Look for an answer choice that is associated with these key words.

© Harcourt

4. A warm front is the border at which a warm air mass moves over a cold air mass. How is a warm front shown on a weather map?

(A) blue lines with half circles

(B) blue lines with triangles

(C) red lines with half circles

(D) red lines with triangles

---

5. Ronald set up a weather station near his house, including the instrument shown here.

Tip
Look for clues on the instrument that indicate its use.

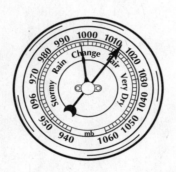

What is the name of this instrument? What does it measure?

_____

_____

Ronald notices a rapid drop in the instrument's reading. What kind of weather is likely to follow?

_____

_____

_____

© Harcourt

1. The ocean floor, shown below, slopes down from the land in different stages.

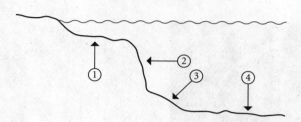

Which of the features shown is the continental rise?

Ⓐ Feature 1

Ⓑ Feature 2

Ⓒ Feature 3

Ⓓ Feature 4

2. An area where a river flows into an ocean is rich in plant and animal life. What is this type of area called?

Ⓐ estuary

Ⓑ jetty

Ⓒ sea arch

Ⓓ tide pool

© Harcourt

3. The illustration below shows a ring of coral that once surrounded a volcanic island. Over time, the island sank and was worn away. The coral remained and, gradually, a grass-covered circular island formed.

Tip
Try to eliminate terms that do not refer to an island.

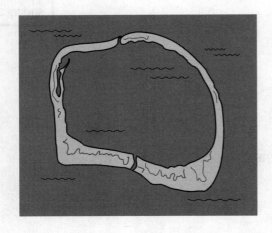

What is the name given to an island that formed this way?

Ⓐ atoll

Ⓑ headland

Ⓒ sea arch

Ⓓ tide pool

4. You are swimming at the seashore when suddenly a rip current begins pulling on you. What should you do?

Ⓐ Swim directly toward the shore as quickly as possible.

Ⓑ Swim parallel to the shore until you leave the current.

Ⓒ Stop swimming and let the current move you toward shore.

Ⓓ Swim away from shore to escape the current.

5. The illustration below shows how the ocean tides are affected by a certain position of the sun, moon, and Earth. The arrows show the pull of the sun and the moon.

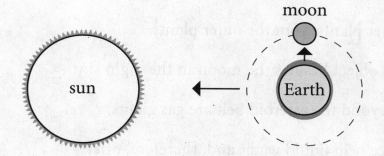

What type of tide is shown in the illustration?

_____

_____

Explain how the level of ocean water is affected during this type of tide.

_____

_____

_____

_____

Name _____

Date _____

1. Max wants to make a model of the solar system for his room at home. He knows the asteroid belt should be included, but he isn't sure why it is important. What would you tell him?

   Ⓐ It divides the inner plants from the outer plants.

   Ⓑ It is the brightest object besides the moon in the night sky.

   Ⓒ All the planets beyond the asteroid belt are gas giants.

   Ⓓ The asteroids once held liquid water and, therefore, offer important clues to how life began on Earth.

2. The illustration shows the moon revolving around Earth, and Earth revolving around the sun. Both the moon and Earth rotate on their axes.

Tip
Apply each answer choice to the illustration to see which would result in the same side of the moon always facing Earth.

   What additional information is needed to explain why the same side of the moon always faces Earth?

   Ⓐ The time for one revolution of the moon equals the time for one rotation of Earth.

   Ⓑ The time for one revolution of the moon equals the time for one revolution of Earth.

   Ⓒ The time for one rotation of the moon equals the time for one rotation of Earth.

   Ⓓ The time for one rotation of the moon equals the time for one revolution of the moon.

© Harcourt

3. Why is it generally warmer in summer than in winter in North America?

Tip
Remember how the tilt of Earth on its axis causes seasonal changes.

Ⓐ Earth's axis always points toward the North Star.

Ⓑ The sun's rays strike more directly during the summer than during the winter.

Ⓒ Earth is farther from the sun during the winter and closer in the summer.

Ⓓ The North Pole is dark all day in the winter and light all day in the summer.

4. Mario uses a softball to model moon phases in his darkened classroom by spinning around on a swivel chair placed next to a lamp. At times, Mario's body is between the lamp and the softball. At other times, the softball is between Mario and the lamp. Which of the following is true?

Ⓐ Mario's arm represents the inner planets.

Ⓑ Mario's spinning represents Earth's orbit around the sun.

Ⓒ Mario represents the sun, the chair represents Earth, and the softball represents the moon.

Ⓓ Mario represents Earth, the softball represents the moon, and the lamp represents the sun.

5. Study the arrangement of Earth, the moon, and the sun shown in the diagram below.

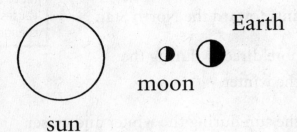

Name the phase of the moon illustrated in the diagram. What would this phase look like from Earth?

_____

_____

What will the phase of the moon look like from Earth about one week later?

_____

_____

_____

Name _____

Date _____

1.

54
Xe
Xenon

Tip
Think about what is true about all xenon atoms.

The periodic table is a collection of boxes, like the one shown above, that show properties of the different elements. What information does this box give about a xenon atom?

Ⓐ It has 54 neutrons.

Ⓑ It has 54 protons.

Ⓒ It has a total of 54 neutrons and electrons.

Ⓓ It has a total of 54 protons and electrons.

2. Where on the periodic table are metals located?

Ⓐ bottom

Ⓑ left

Ⓒ right

Ⓓ top

© Harcourt

3. In some northern locations, snow will change to water vapor in the air, even when the temperature is below the freezing point. What happens to the snow in this case?

> **Tip**
> Think about how each change occurs, and eliminate answer choices that you know can't occur below the freezing point of water.

   (A) It melts.

   (B) It evaporates.

   (C) It condenses.

   (D) It sublimes.

4. Which of the following is a chemical change?

   (A) rusting of a metal

   (B) freezing of water

   (C) evaporation of a liquid

   (D) breaking of glass

5. The table below gives the mass and volume of blocks of different metals.

Tip
Density is the mass of a substance divided by its volume.

Metal	Mass (g)	Volume (cm³)
Aluminum	8.1	3.0
Gallium	16.3	2.76
Gold	11.0	1.4
Nickel	18.7	2.1

According to the table, which metal has a density of 8.9 g/cm³?

_____

If you heated the metal objects, their volumes would increase. How would this affect their densities? Explain.

_____

_____

_____

Name _____

Date _____

1. The table gives the melting points of different solids.

Substance	Melting point (°C)
Aluminum	660
Gold	1064
Lead	327
Potassium	64
Sodium	98

> **Tip**
> Think about how the melting point of a substance is related to the amount of energy needed to separate its particles.

Based on the information in the table, which of the following is a true statement about the amount of energy needed to separate particles in the solids?

(A) More energy is needed for gold than for aluminum.

(B) More energy is needed for potassium than for sodium.

(C) More energy is needed for lead than for gold.

(D) More energy is needed for sodium than for lead.

2. Which is an example of heat transfer *only* by conduction?

(A) Rays from the sun strike a beach.

(B) Heat moves from a floor heating vent toward the ceiling.

(C) A fire burning in a fireplace warms a room.

(D) Your hand becomes warm when you hold a cup of hot chocolate.

3.

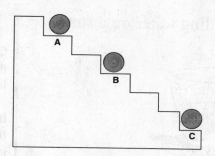

The picture shows different positions of a ball as it starts from rest at point A, rolls down the stairs, and stops at point C before reaching the floor. Which is a true statement about the energy of the ball as it rolls down the stairs?

Ⓐ The ball has only kinetic energy at point B.

Ⓑ The ball has only kinetic energy at point A.

Ⓒ The ball has only potential energy at point C.

Ⓓ The ball has only potential energy at point B.

Tip
Think about whether the ball is moving or at rest at each point.

---

4. How is radiation different from other forms of heat transfer?

Ⓐ It can only occur between objects at different temperatures.

Ⓑ It doesn't need matter for the heat to travel through.

Ⓒ It only occurs when the warmer object is burning.

Ⓓ It can occur in empty space but not through matter.

© Harcourt

Name _____ Date _____

5. The illustration shows a pot of boiling water on a stove.

> **Tip**
> Consider whether
> thermal energy moves
> from one object to
> another, or whether
> thermal energy moves
> through the
> movement of a gas or
> a liquid.

Describe how thermal energy moves by conduction as
the water boils.

_____

_____

_____

Describe how thermal energy moves by convection as
the water boils.

_____

_____

_____

1. What is the biggest advantage of using electromagnets, instead of regular magnets, in electric motors?

   Ⓐ Electromagnets are stronger.

   Ⓑ Less energy is used by electromagnets.

   Ⓒ Electromagnets can be turned on and off easily.

   Ⓓ Regular magnets cost more than electromagnets.

2.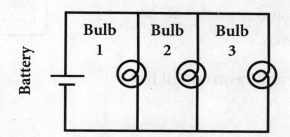

   Tip
   Consider how current moves through the circuit.

   What would happen if Bulb 1 were removed from this parallel circuit?

   Ⓐ Bulbs 2 and 3 would stay lit.

   Ⓑ Bulbs 2 and 3 would go out.

   Ⓒ Bulb 2 would stay lit; bulb 3 would go out.

   Ⓓ Bulb 3 would stay lit; bulb 2 would go out.

© Harcourt

3. The light bulb a student uses in a desk lamp is labeled "60 watts." What property does the label refer to?

(A) current per second

(B) energy per second

(C) resistance per second

(D) voltage per second

4. What is always true about a neutral atom?

Tip
Think about what the term *neutral* means.

(A) The protons, electrons, and neutrons all have the same, non-zero charge.

(B) The protons, electrons, and neutrons all have zero charge.

(C) The total charge of the protons is the same as the total charge of the electrons.

(D) The total charge of the protons is opposite the total charge of the electrons.

© Harcourt

Name _____ Date _____

5. Keisha built this simple circuit with a battery, light bulb, switch, and wires.

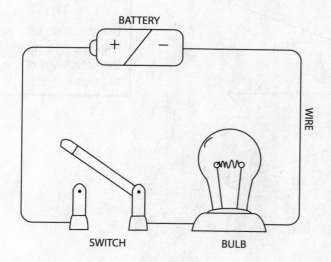

Compare the flow of electricity through the circuit when the switch is open and when it is closed.

_____

_____

_____

_____

Identify the changes in forms of energy when electricity flows through the circuit. Consider electrical, chemical, thermal, light, and other forms of energy.

_____

_____

_____

_____

© Harcourt

1.

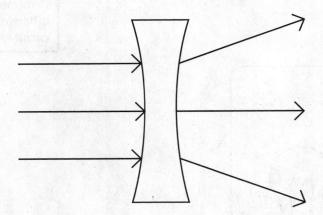

Which of the following statements describes what is happening in the illustration?

Ⓐ Light is reflected as it moves through a concave lens.

Ⓑ Light is reflected as it moves through a convex lens.

Ⓒ Light is refracted as it moves through a concave lens.

Ⓓ Light is refracted as it moves through a convex lens.

2. In which of the following materials would you expect sound to travel the slowest?

Ⓐ warm granite

Ⓑ cold granite

Ⓒ warm air

Ⓓ cold air

> **Tip**
> Remember that sound travels as energy moves from particle to particle. Think about how the type of material and its temperature would affect this process.

---

3. A flute has a higher pitch than a tuba. What else must the sound waves made by a flute have when compared to those made by the tuba?

Ⓐ lower volume

Ⓑ higher frequency

Ⓒ fewer vibrations

Ⓓ lower energy

© Harcourt

4. Which term describes light waves?

    Ⓐ compressional

    Ⓑ constructive

    Ⓒ translucent

    Ⓓ transverse

---

5. Look at the simplified version of the electromagnetic spectrum shown below. The frequencies are not drawn to scale, but the illustration shows that the frequency of radio waves is different from the frequency of X rays.

> **Tip**
> Think about the relationship between the frequency and energy of waves that make up the electromagnetic spectrum.

Radio waves   X rays

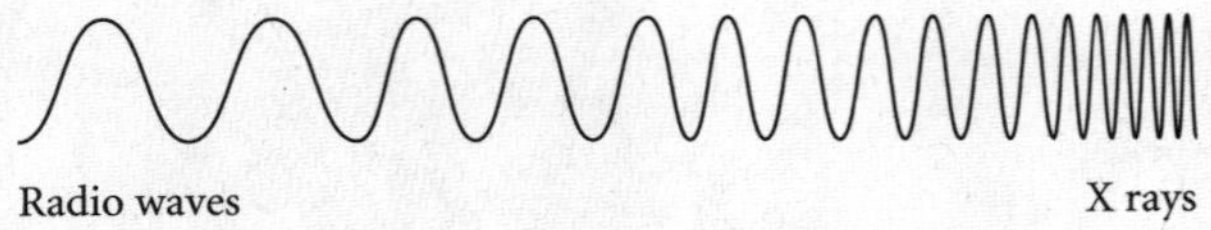

Think about the properties of radio waves and X rays. Which has higher frequency? Which has higher energy?

_____

_____

If someone broke an arm, why would a physician use X rays rather than radio waves to make a film image of the broken bone?

_____

_____

_____

1. What force opposes motion?

    (A) buoyancy

    (B) friction

    (C) gravity

    (D) magnetism

> **Tip**
> The key words in the question are *opposes motion.*

2. Which types of forces cancel out each other?

    (A) balanced forces that are of equal strength

    (B) balanced forces that are **not** of equal strength

    (C) unbalanced forces that are of equal strength

    (D) net forces that are **not** of equal strength

© Harcourt

3.

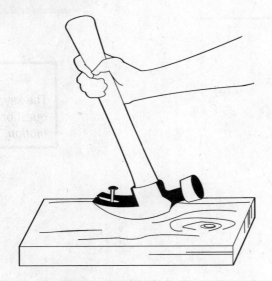

Tip
Where does the effort to pull the nail out of the board come from?

The hammer shown above is an example of a lever.
Where is the *effort force* in the illustration?

Ⓐ the board pushing against the head of the hammer

Ⓑ the hand pulling on the handle of the hammer

Ⓒ the head of the hammer pulling on the nail

Ⓓ the head of the hammer pushing on the board

4. Which device is a compound machine?

Ⓐ bicycle

Ⓑ ramp

Ⓒ seesaw

Ⓓ wedge

© Harcourt

Name _____ Date _____

5. Look at the two machines shown in the illustrations below.

Tip
Think about the different types of simple machines. Recall the main parts of a lever.

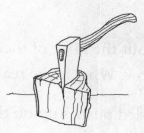

Which two simple machines make up the ax?

_____

_____

Which part of the wheelbarrow is used as a lever system? Where is the fulcrum of the lever system?

_____

_____

_____

© Harcourt

1. The girl in the illustration is riding a sled down a snow-covered hill.

Tip
Apply Newton's third law of motion to the illustration.

Think about the force of the sled pushing down on the snow. What is the reaction force?

Ⓐ The sled pushes up on the girl

Ⓑ Gravity pulls the sled down the hill.

Ⓒ The snow pushes up on the sled.

Ⓓ The air pushes back against the sled.

2. A bicyclist is traveling along a level path at a speed of 2.5 m/sec. The bicyclist then starts down a tall hill and accelerates at a rate of 0.15 m/sec². What is the speed of the bicyclist after accelerating for 4 seconds?

Ⓐ 2.65 m/sec

Ⓑ 2.80 m/sec

Ⓒ 3.10 m/sec

Ⓓ 6.65 m/sec

© Harcourt

3.

Time (s)	Distance (m)
1	9.8
2	19.6
3	29.4
4	39.2

Tip
Look for a pattern in the boulder's velocity during each second.

A boulder falls from a high cliff toward the ground. The table shows the total distance it falls during each second for the first four seconds. What is the boulder's acceleration?

Ⓐ 4.9 m/sec²

Ⓑ 9.8 m/sec²

Ⓒ 19.6 m/sec²

Ⓓ 29.4 m/sec²

4. Think about the different forces that act on an airplane in flight. What is the forward force produced by the plane's engine?

Ⓐ drag

Ⓑ friction

Ⓒ lift

Ⓓ thrust

5. Dwayne builds a model boat that has a mass of 75 g. He sets the boat in the water and points a pocket fan at the sail, causing the boat to accelerate.

Tip
Determine the velocity after 2 seconds, and then use the velocity to calculate the momentum.

The boat is initially at rest, but the fan causes it to accelerate at a rate of 2.5 cm/sec^2. After 2 seconds, what is the momentum of the boat?

_____

_____

_____

Does the momentum of the boat stay the same, or is it changing? Explain.

_____

_____

_____

_____

Name _____

Date _____

1. The periodic table uses symbols to identify common elements. Look at the boxes below from the periodic table.

Which of the symbols represents the element sodium?

Ⓐ Ar

Ⓑ Be

Ⓒ Mg

Ⓓ Na

2. Sandy likes the ferns that are growing in a flowerbed shaded by trees. She digs up the ferns and then plants them next to the mailbox, where it is bright and sunny. She gives the ferns the right amount of water and nutrients (plant food). Two months later, the ferns are dead. Why did the ferns do so poorly after Sandy moved them?

Ⓐ Ferns need shade, not bright sunlight.

Ⓑ They did not get enough water or minerals.

Ⓒ Insects ate the leaves faster than they could grow.

Ⓓ The soil near the mailbox was too rocky for plants to grow.

© Harcourt

3. Which of these is a properly ordered food chain in a forest ecosystem?

Ⓐ bear ⟶ eagle ⟶ fox

Ⓑ fox ⟶ bear ⟶ eagle

Ⓒ grass ⟶ eagle ⟶ mouse

Ⓓ grass ⟶ rabbit ⟶ fox

4. Rita used different colors of clay to make the model below.

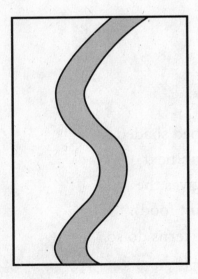

Which of the following was she making a model of?

Ⓐ delta

Ⓑ meander

Ⓒ mesa

Ⓓ spit

5. Beach mice come out to feed at night. A group of four scientists tested the hypothesis that artificial lights cause beach mice to eat less. The scientists went to a beach with no artificial lights. They chose eight similar places on the beach. On a dark night, they set up lights in four places and placed trays of food near each light. They also placed trays of food in four dark places. Why was it important to have four lighted places and four dark places?

　Ⓐ to make sure the mice had enough to eat

　Ⓑ to make sure the whole beach was included in the experiment

　Ⓒ so that the scientists could carry out four separate experiments and not share their findings

　Ⓓ so that a fair comparison could be made between mice in lighted places and mice in dark places

6. Scientists have found that Earth is divided into four layers. The thickest is about 2,885 kilometers deep. Which of the four layers is the thickest?

　Ⓐ crust

　Ⓑ mantle

　Ⓒ outer core

　Ⓓ inner core

7. The drawings show a mountain valley at two different
times. In the first drawing, the valley contains a glacier.
In the second drawing, the glacier has melted.

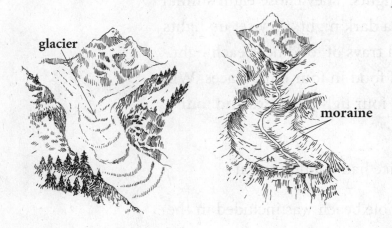

Look at the first drawing. How does the glacier erode
the land beneath it?

_____

_____

_____

Look at the second drawing. What is the moraine made
of, and how was it formed?

_____

_____

_____

_____

8. Derek is doing an investigation with pendulums. He is counting the number of times different pendulums swing in one minute. His data do not support his prediction. What should Derek do?

  (A) Make up data that support his prediction.

  (B) Turn in his report without any recorded data.

  (C) Record the data he observes, and then repeat the investigation.

  (D) Copy the data that his classmates collected.

9. The diagram shows a system for heating a swimming pool by using solar energy.

  In which direction does energy flow in this system?

  (A) sun, pipes, swimming pool, solar panel

  (B) swimming pool, pipes, solar panel, sun

  (C) pipes, swimming pool, solar panel, sun

  (D) sun, solar panel, pipes, swimming pool

10. Diagrams of electrical circuits contain symbols to represent different parts of the circuits. Look at the symbol below.

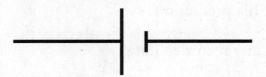

Which part of an electrical circuit does the symbol represent?

(A) battery

(C) lamp

(B) conductor

(D) switch

11. Ming performs an experiment to test the effect of salt water on the chemical makeup of a chicken egg. He places the egg in salt water and each day picks up the egg to see how it has changed. What is the last thing Ming should do each day after he checks the egg?

(A) Rinse the egg.

(B) Put on safety goggles.

(C) Wash his hands.

(D) Dry the egg.

© Harcourt

12. Which of the following would **most likely** cause the fastest erosion?

   Ⓐ light rain falling on a sandy beach

   Ⓑ swift river water moving over rocks

   Ⓒ waves crashing against a sandy shore

   Ⓓ rain running down a grass-covered hill

13. Mr. Martin's class is investigating cloud formation. They set up two containers, as shown. They place both containers in full sunlight. On the lid of one container, they place an ice cube.

Inside which container will water droplets form first, and why?

   Ⓐ the one with the ice, because the temperature is lower

   Ⓑ the one without the ice, because the temperature is higher

   Ⓒ the one with the ice, because the melting ice leaks inside the container

   Ⓓ the one without the ice, because it contains air with more water vapor

© Harcourt

14. Erica compared the mass of iron before and after it had rusted. She measured two piles of iron shavings and found that they had the same mass. She kept one pile protected, but she allowed the other pile to rust. Afterwards, Erica placed the piles on a balance. To balance the piles, she had to add standard masses to the side with the non-rusted pile.

iron shavings   **rusted
iron shavings**

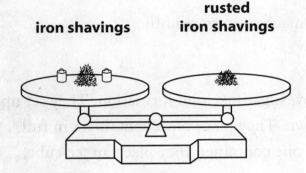

Which of the following explains what happened?

Ⓐ Some of the iron's mass was destroyed during the rusting process.

Ⓑ Erica measured the masses incorrectly before the iron rusted.

Ⓒ Another substance combined with the iron when it rusted.

Ⓓ Mass was created and added to the iron during the rusting process.

15. Why is an electromagnet more useful in a doorbell than a permanent magnet would be?

   Ⓐ An electromagnetic can make the sound louder.

   Ⓑ An electromagnetic can be turned on and off.

   Ⓒ A permanent magnet would be too weak.

   Ⓓ A permanent magnet would be too heavy.

16. The carrot and the barrel cactus are both flowering plants. The carrot has one large, deep taproot that stores food for the plant. Its very short stem holds up many long, feathery leaves. The barrel cactus has a small taproot and long, thin, shallow roots that spread out in all directions. Its stems grow together to form a mass that stores water. The barrel cactus has spines instead of leaves.

   Name a way that the barrel cactus and the carrot are alike.

   _____

   _____

   _____

   How are the roots of the barrel cactus and the carrot different?

   _____

   _____

   _____

© Harcourt

17. A student wants to perform an experiment to test how much water bean plants need for good growth. Which variable should be changed?

Ⓐ the type of plant

Ⓑ the amount of sunlight

Ⓒ the type of soil

Ⓓ the amount of water

18. The figures below show the area where a river flows into the ocean. The figure on the right shows what the area looks like two years after the figure on the left.

Before

After

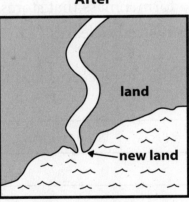

Notice that land has built up near the mouth of the river where the water has deposited rocks and soil. What is the buildup of land called?

Ⓐ delta

Ⓒ floodplain

Ⓑ gully

Ⓓ landslide

19. Look at the pond food chain below.

pond weed ⟶ insects ⟶ frogs ⟶ snakes

If a large number of frogs were captured and removed from the pond, what would **most likely** happen?

Ⓐ Many insects would die, and the number of snakes would decrease.

Ⓑ The number of insects would increase, and many snakes would die.

Ⓒ The number of insects would decrease, and the number of snakes would decrease.

Ⓓ The number of insects would increase, and the number of snakes would remain about the same.

20. Students in a class plan to measure the humidity outside their school each day for a week. What instrument should they use?

Ⓐ anemometer

Ⓑ barometer

Ⓒ hygrometer

Ⓓ thermometer

© Harcourt

21. Dylan conducted an experiment to compare the densities of different liquids. He carefully poured two liquids at a time into a graduated cylinder and observed which floated and which sank. The figure below shows his observations.

Based on his observations, which liquid can Dylan conclude has the **lowest** density?

Ⓐ corn oil

Ⓒ maple syrup

Ⓑ dish soap

Ⓓ water

22. Nonrenewable resources should be used carefully, because they cannot be replaced within a human lifetime. Which of the following is an example of a nonrenewable resource?

Ⓐ plants

Ⓒ wind

Ⓑ minerals

Ⓓ sunlight

Name _____

Date _____

23. The graph shows densities for different types of rocks.

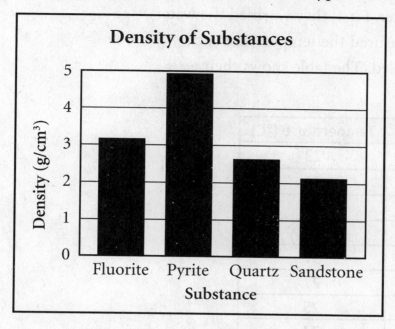

Limestone has a density of 2.68. Which of the substances listed in the graph has a density **closest** to that of limestone?

(A) Fluorite   (C) Quartz

(B) Pyrite   (D) Sandstone

24. Scientists have found that fossils and rock formations along the eastern coast of South America are similar to those along the western coast of Africa. What theory does this support?

(A) continental drift

(B) fossil formation

(C) landform erosion

(D) secondary succession

© Harcourt

25. Some students performed an experiment to determine the cooling time for water. First, they warmed the water to 37°C. Then, they measured the temperature each minute as the water cooled. The table shows their measurements.

Time (min)	Temperature (°C)
0	23
I	31
2	37
3	34
4	31
5	27
6	25
7	24
8	23

How many minutes after the water reached 37°C did it take to cool back to its starting temperature?

Ⓐ 2

Ⓑ 4

Ⓒ 6

Ⓓ 8

© Harcourt

26. Suppose a population of fish lives in a small pond. The gills of some of the fish are better at obtaining oxygen than are the gills of other fish in the population. The temperature of the pond water rises over many years. Warmer water holds less oxygen than cooler water. What may happen to the fish as the temperature of the pond rises?

 Ⓐ Some fish will move to a new, cooler pond.

 Ⓑ Only fish with more efficient gills may survive.

 Ⓒ All of the fish will die from lack of oxygen.

 Ⓓ The fish will share the oxygen, and will all survive.

27. Ginny uses small colored paper circles to model elements and compounds. Which of the following is the **best** model to use for an element?

 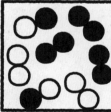 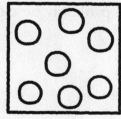

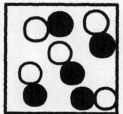 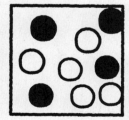

© Harcourt

28. Wesley measures the daily high temperature at his home for the first 15 days of the month. The graph below shows his measurements.

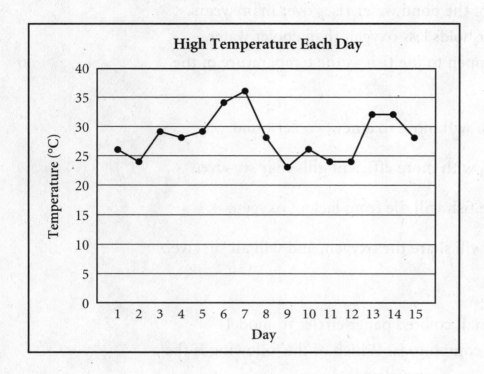

If he measures the temperature again the next day, which of the following will probably be **closest** to his measurement?

Ⓐ 5°C

Ⓑ 15°C

Ⓒ 25°C

Ⓓ 35°C

29. The illustration below shows a type of electrical circuit that was once used for strings of decorative lights.

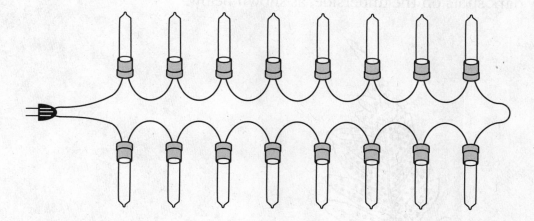

What is this type of circuit called, in which the current has only one path to follow?

_____

What are two disadvantages of using this type of circuit?

_____

_____

_____

© Harcourt

30. Students in a class were asked to bring in leaves to study. Ann brought in a fern frond, but when she turned it over, she saw dark spots on the underside, as shown below.

What do the dark spots show about how this fern will reproduce?

Ⓐ Part of the fern will break off to form a new plant.

Ⓑ A single cell will grow into a new fern plant.

Ⓒ The fern will reproduce by cell division.

Ⓓ Two gametes will join to form a zygote.

© Harcourt

31. The Miller family was proud of their lawn. They used fertilizer to keep it thick and green, herbicide to kill weeds, and pesticides to kill insects. Then the Millers moved away, and the Schwartz family moved in. The Schwartzes did not use chemicals. After a few years, the lawn was thinner. In some places insects in the soil had killed the grass. However, many beautiful birds now visit the yard. The birds were not there before. What happened?

Ⓐ The herbicides stopped killing the grass, so it grew thicker.

Ⓑ Without fertilizer, the lawn was thicker and more beautiful.

Ⓒ Without pesticides, insects came back, and so did birds that eat them.

Ⓓ The birds were able to see the insects better through the thinner grass.

32. The figure below shows a ball rolling down a hill, across a flat space, and up another hill.

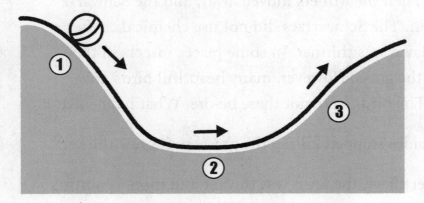

Describe how gravity affects the ball's speed at each of the three points labeled on the figure.

_____

_____

_____

Describe how friction affects the ball's speed at each of the three points.

_____

_____

1. Solar energy is a renewable resource. Clouds reflect solar energy back to space. The table below shows the average number of cloud-free days each year in four cities.

Cloud-Free Days per Year	
City	Cloud-Free Days (average)
Charleston, South Carolina	101
Houston, Texas	94
Phoenix, Arizona	214
Portland, Oregon	69

In which city would it make the most sense to use solar energy?

(A) Charleston, South Carolina

(B) Houston, Texas

(C) Phoenix, Arizona

(D) Portland, Oregon

2. A speaker in a stereo system uses an electromagnet to make parts inside the speaker vibrate. Why is an electromagnet used rather than a permanent magnet?

(A) It can make the cone vibrate faster.

(B) It can be made into larger sizes.

(C) It has a stronger pull on the cone.

(D) It can be turned on and off.

© Harcourt

3. Tom is doing an experiment. His hypothesis is that water runs through gravel faster than it runs through sand or clay. The diagram shows his experimental setup. He plans to pour equal amounts of water into the top of each container and then time how long it takes for water to begin dripping from the bottom of each. Tom puts the same amount of soil in each container. The soil is held in the containers by identical squares of cloth kept in place by rubber bands. The containers are fastened to ring stands.

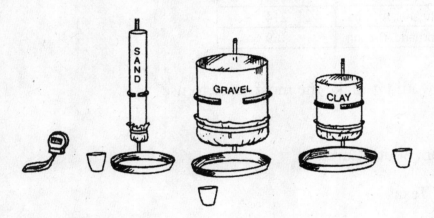

What is the tested variable in Tom's experiment?

_____

How can Tom modify his setup to make certain that the tested variable is the only variable that changes?

_____

_____

_____

© Harcourt

4. New roads, buildings, and parking lots break up large areas of habitat into smaller pieces. A group of scientists is investigating how breaking up habitat affects a certain type of bird. The table below shows data from the investigation. The area of each habitat is given in hectares (ha), a metric unit of area. (1 ha = 2.47 acres)

Nests in Habitat Areas		
Habitat Location	Area of Habitat (ha)	Number of Nests in Habitat
A	2.0	12
B	0.25	0
C	0.75	3
D	2.1	11
E	0.5	1

Which hypothesis are the scientists testing?

(A) Polluted habitats will have fewer nests.

(B) Habitats with less area will have fewer nests.

(C) Habitats near highways will have fewer nests.

(D) Habitats near bird feeders will have more nests.

5. Carlos holds a tennis ball (mass = 57 g) and a golf ball (mass = 46 g) a distance of 2 meters above the ground. He then releases them and observes the height of their first bounces. Carlos notices that the golf ball bounces higher, but neither ball bounces back up to the 2-meter level. What can Carlos conclude?

Ⓐ Both balls have more energy bouncing up than they had falling.

Ⓑ The tennis ball had a lower bounce because it is lighter.

Ⓒ The tennis ball started out with less energy.

Ⓓ The golf ball loses less of its energy to the ground.

6. The weather forecast is for warmer air and lower air pressure. What other change would you expect?

Ⓐ higher humidity

Ⓑ strong winds

Ⓒ lower humidity

Ⓓ violent winds

© Harcourt

7. Look at the simple electrical circuit diagram shown below.

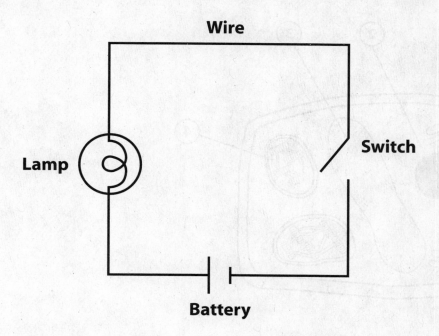

Which part provides electrical pressure in the circuit?

(A) battery

(B) lamp

(C) switch

(D) wire

8. Different organelles are labeled on the following plant cell.

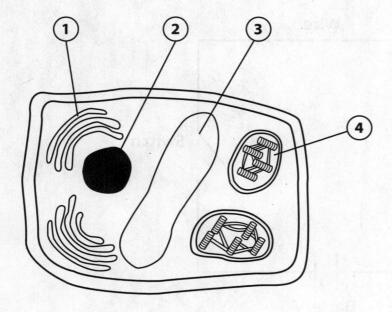

Which organelle is a vacuole?

(A) 1

(B) 2

(C) 3

(D) 4

9. In the early 1990s, an experiment tested whether humans could survive inside an artificial Earth environment, called Biosphere 2. The structure contains five different ecological communities and a small ocean. Eight people spent two years inside Biosphere 2. They were completely closed in, with everything they needed for survival. Why did Biosphere 2 have to include plants?

   Ⓐ The plants shaded the people from the sun.

   Ⓑ The plants provided fresh water for the people.

   Ⓒ The plants kept the people from getting homesick.

   Ⓓ The plants provided the people with food and oxygen.

10. Felicia is using clay to make a model of different types of volcanoes. How should she form the clay if she wants to make a model of a cinder cone volcano?

   Ⓐ tall, narrow, steep slopes

   Ⓑ wide, fairly steep slopes

   Ⓒ wide, fairly gentle slopes

   Ⓓ short, narrow, gentle slopes

© Harcourt

11. Symbiosis is a relationship between different types of organisms. The table describes different kinds of symbiosis.

Types of Symbiosis	
Mutualism	Both organisms benefit.
Commensalism	One organism benefits, and the other isn't affected.
Parasitism	One organism benefits, but the other is harmed.

Which of the following is an example of mutualism?

(A) A flea takes in the blood of a dog. The bite of the flea itches the dog, and the dog may get a disease.

(B) Barnacles on a whale eat food scraps from the whale. The whale is not affected.

(C) Bees drink nectar from flowers. Flowers can reproduce because bees carry pollen between them.

(D) A hermit crab lives in the empty sea shell of a snail. The crab uses the shell for protection.

12. A class uses an anemometer each day for a week to make measurements. What are they measuring?

(A) air pressure

(B) humidity

(C) wind chill factor

(D) wind speed

13. The diagrams below show the side of a cliff. The one on the right shows the cliff six weeks later than the one on the left.

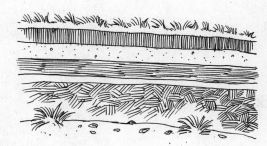 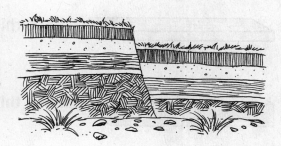

What is the most likely cause for the sudden change in the cliff, and why did the change take place where it did?

_____

_____

_____

What kind of landform will result if the same thing happens over and over again?

_____

© Harcourt

14. Students in a class are testing how sound moves through different wooden boards. The figures below show the shapes they test. All of the boards are made of the same type of wood, but they have different widths and lengths. The students support each board at its ends, and observe differences in sound when they hit the center of each board with a stick.

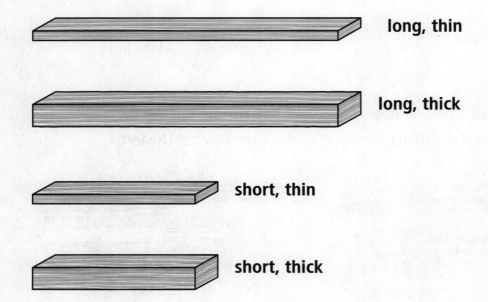

long, thin

long, thick

short, thin

short, thick

Which of the boards **most likely** produced the sound waves with the **highest** frequency?

(A) long, thin

(B) long, thick

(C) short, thin

(D) short, thick

© Harcourt

15. The periodic table is a collection of details about chemical elements. Each element has a symbol that stands for the element. Which of the following elements has an incorrect symbol written on its box?

Ⓐ
```
   6
   C
 Carbon
```

Ⓒ
```
   12
   Mg
Magnesium
```

Ⓑ
```
   7
   Ni
 Nitrogen
```

Ⓓ
```
   13
   Al
 Aluminum
```

16. Why should scientists accurately report the results of an experiment?

Ⓐ Other scientists will be able to graph the data that they report.

Ⓑ The work that the scientists have done can be admired by other scientists.

Ⓒ The results of the experiment can be disproved by other scientists.

Ⓓ Others can repeat the experiment to see if they get similar results.

© Harcourt

17. Farmers can grow more food with fertilizer than they can without. More food helps keep prices low, and lower prices mean that fewer people go hungry. However, fertilizer runs off fields and into ponds and streams, where it causes too much algae to grow. When the algae die, decomposers use up the oxygen in the water as they break down the dead algae. Based on this information, what is a **likely** result of fertilizer use?

(A) Food prices will rise.

(B) Fish will eat more algae.

(C) Pond water will be clearer.

(D) More fish will die than usual.

18. Why are index fossils helpful for determining the age of a rock layer that has been moved out of its normal position by an earthquake?

(A) They formed at only one particular time.

(B) They are the oldest fossils ever found.

(C) They show the relative age of rock formations.

(D) They are evidence of animal activities in an area.

© Harcourt

19. Mr. Johnson's class is investigating the water cycle. The students use the experimental setup shown below. They record information about the level of water in the smaller cup and about the number of droplets that condense on the lid and sides of the larger cup.

What is Mr. Johnson's class doing?

Ⓐ analyzing data

Ⓑ observing a model

Ⓒ drawing a conclusion

Ⓓ forming a hypothesis

20. The table below shows the densities of different types of wood.

Densities of Wood	
Type of Wood	Density (g/cm³)
Beech	0.70
Ebony	1.11
Hickory	0.93
Juniper	0.56

If a small block of each type of wood is placed in a beaker filled with water, which type of wood will sink?

(A) beech

(B) ebony

(C) hickory

(D) juniper

21. Kevin's nervous system helps him to start running toward a friend he sees. What other body systems work together to help Kevin move?

(A) circulatory and excretory

(B) digestive and skeletal

(C) muscular and skeletal

(D) respiratory and digestive

22. The eggs of some plants can only be fertilized in certain environments. For which of the following plants are eggs fertilized in water?

   Ⓐ ferns and mosses

   Ⓑ ferns and gymnosperms

   Ⓒ mosses and angiosperms

   Ⓓ angiosperms and gymnosperms

23. Peter brings a liquid to a boil and then measures the time for it to cool back to room temperature (22°C). The graph below shows his measurements.

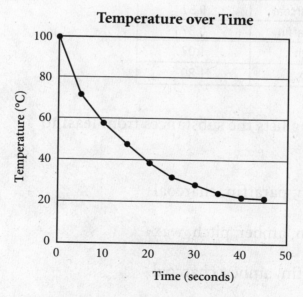

About how long did it take the liquid to cool to 40°C?

   Ⓐ 10 seconds   Ⓒ 22 seconds

   Ⓑ 18 seconds   Ⓓ 27 seconds

© Harcourt

24. In the 1600s, Robert Hooke used a microscope to study thin slices of cork. He noticed the cork was divided into tiny, boxlike structures.

What were these boxlike structures that Hooke saw?

_____

Why were Hooke's observations so important?

_____

_____

25. The table lists the densities of some solid substances.

Substance	Density (g/cm³)
Amber	1.06
Charcoal	0.57
Paraffin	0.87
Pitch	1.07
Wax	1.80

Which of the following lists the substances from **least** to **greatest** density?

Ⓐ wax, amber, pitch, paraffin, charcoal

Ⓑ charcoal, paraffin, amber, pitch, wax

Ⓒ wax, pitch, paraffin, amber, charcoal

Ⓓ charcoal, amber, pitch, paraffin, wax

26. A common light bulb used in overhead lamps has "100 watts" written on it. The term "watt" describes what quantity each second?

   Ⓐ energy

   Ⓑ current

   Ⓒ electric charge

   Ⓓ electrical pressure

27. A scientist conducts an experiment to test a theory that is accepted by other scientists as being true. The results of the experiment do not agree with the theory. What should the scientist do?

   Ⓐ Change the results to match the theory, because the theory is known to be true.

   Ⓑ Realize that the theory may be wrong, and plan experiments to further test it.

   Ⓒ Assume that the results of the experiment are incorrect, and do not report them.

   Ⓓ Determine what mistake was made in the experiment to cause the incorrect results.

© Harcourt

28. In science class, Jen is making a model of the protons, electrons, and neutrons in atoms using gum drops. Which of the following breaks a safety rule in science class?

    Ⓐ gluing the gum drops to poster board

    Ⓑ allowing other students to use the gum drops

    Ⓒ cutting some of the gum drops in half

    Ⓓ eating some of the gum drops used in the model

29. In order for ice skates to work properly, they must be able to slide easily over the ice.

    Describe ways in which friction affects the motion of an ice skater.

    _____

    _____

    Why would skaters sharpen the blades of their skates before a practice or performance?

    _____

    _____

    _____

© Harcourt

30. The graph shows how air pressure changes with height above Earth (altitude).

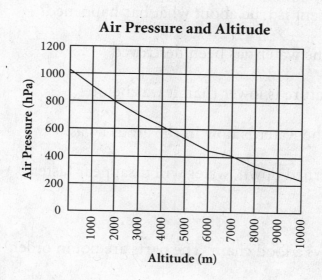

Air Pressure and Altitude

Infer the air pressure at an altitude of 11,000 meters.

Ⓐ about 100 hPa

Ⓑ about 200 hPa

Ⓒ about 300 hPa

Ⓓ about 400 hPa

31. Alberto's uncle boils water on the stove to add moisture to the air. The pot starts out full. After twenty minutes, the pot is only half full. Which statement is true about what has happened?

Ⓐ Half the mass of the water has been destroyed.

Ⓑ The water temperature is lower than it was before.

Ⓒ Half the mass of the water is now in the air as a gas.

Ⓓ If the burner is turned down, water will disappear faster.

32. The picture below shows a food chain. The parts are not in order.

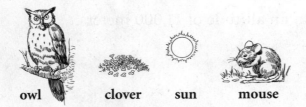

owl   clover   sun   mouse

Which list gives the correct order in which energy flows through this food chain?

Ⓐ owl, mouse, clover, sun

Ⓑ clover, mouse, owl, sun

Ⓒ sun, clover, mouse, owl

Ⓓ mouse, clover, owl, sun

Name _____

Date _____

1. A class is using small colored candy pieces to model the differences in solids, liquids, and gases. What important safety procedure should the students follow?

   Ⓐ Do not allow any other student to touch the candy pieces.

   Ⓑ Always wear rubber gloves when working with the candy.

   Ⓒ Do not eat any of the candy when the activity is finished.

   Ⓓ Tell the teacher if any of the candy pieces falls to the floor.

2. The diagram below shows a simple electrical circuit.

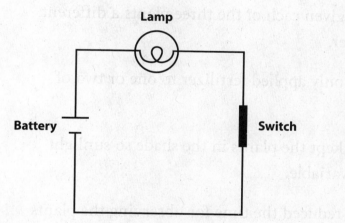

   Which of the following would cause the lamp in the circuit to glow brighter?

   Ⓐ Add another battery to the circuit.

   Ⓑ Add another lamp in series.

   Ⓒ Decrease the voltage of the battery.

   Ⓓ Open the switch in the circuit.

3. Tae conducts an experiment to test the effect of fertilizer on plant growth. The steps of his experiment are listed below.

> 1. Obtain three plants.
> 2. Observe the plants for two weeks. Give each plant the same amount of water and sunlight.
> 3. Apply fertilizer to each of the plants.
> 4. Observe the growth of the plants for two more weeks.

Tae concludes that fertilizer improves plant growth. What could he have done to make his conclusion more reliable?

(A) He could have given each of the three plants a different amount of water.

(B) He could have only applied fertilizer to one or two of the plants.

(C) He could have kept the plants in the shade so sunlight wouldn't be a variable.

(D) He could have reduced the time for observing the plants before applying fertilizer.

4. Yasmin is testing the hypothesis that if a mouse runs a maze more than once, it will run faster each time. Her procedure and data are listed below.

Procedure

1. Start investigation at feeding time so mouse is hungry.

2. Place food at end of maze. Place mouse at start of maze.

3. Time how long it takes mouse to find food (in seconds).

4. Let mouse eat food, and then remove mouse.

5. Repeat steps 2 through 4 until eight trials have been run.

Trial	Time (Sec)
1	170
2	90
3	45
4	10
5	10
6	10
7	115
8	*

*In Trial 8, mouse wandered around and then climbed out.

Explain whether Yasmin's data support her hypothesis.

_____

_____

Review the procedure and the data. How could Yasmin modify the procedure so that she could have greater confidence in the results?

_____

_____

_____

_____

© Harcourt

5. The table shows the dates of moon phases for four months.

Moon Phases—Summer 2005				
Month	New Moon	First Quarter	Full Moon	Third Quarter
June	6	15	22	28
July	6	14	21	28
August	5	13	19	26
September	4	11	18	25

Predict the date of the next full moon.

(A) October 2

(B) October 10

(C) October 17

(D) October 24

6. Which of the following causes seasonal changes on Earth?

(A) Earth's orbit and tilt

(B) the shape of the sun's orbit

(C) the changing shape of Earth's orbit

(D) Earth's position relative to the moon

© Harcourt

7. The diagram shows a mountain with a cloud near its peak.

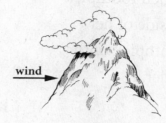

wind

Based on your understanding of how landforms affect the water cycle, how did the cloud get there?

Ⓐ It formed high over the ocean and was moved by wind to the mountain.

Ⓑ It formed in a dry area above the mountain and was pushed down by an air current.

Ⓒ It formed as fog over the ocean, and the wind pushed it up the side of the mountain.

Ⓓ It formed there when humid air was pushed up the mountainside and cooled off.

8. Loren is testing the solubility of sugar in water. As he stirs sugar into a glass beaker of water, he accidentally drops the beaker. The glass breaks on the table, spilling the sugar water. What should he do now?

Ⓐ Use thick paper towels to mop up the mess.

Ⓑ Ask another student to help him clean up.

Ⓒ Tell his teacher about the accident and mess.

Ⓓ Sweep the glass and sugar water into a garbage can.

© Harcourt

9. Daria is going to observe some cells under a microscope. She uses a flat wooden toothpick to scrape some cells from the inside of her mouth. Then she scrapes the toothpick onto a slide and looks at the slide under a microscope.

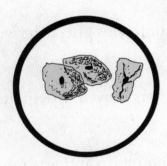

How does she know that these cells are not from the wooden toothpick?

Ⓐ They are too orderly.

Ⓑ They do not have cell walls.

Ⓒ They do not have a nucleus.

Ⓓ There are too many of them.

© Harcourt

10. Carl is writing a lab report for an experiment. His results do not support the hypothesis he was testing. What should Carl do?

Ⓐ Change his hypothesis to fit his results.

Ⓑ Call his friends and use their results instead.

Ⓒ Change his results so they support his hypothesis.

Ⓓ Write that the results do not support the hypothesis.

11. Elevators use a system of pulleys and weights to raise and lower loads. Before elevators were invented, almost everyone had to walk or be carried up and down stairs. Elevators have given greater freedom to people who cannot walk. But because of elevators, people who can walk climb stairs less than they did before. Which is the **most likely** result of the invention of the elevator?

Ⓐ People get less exercise.

Ⓑ Shopping malls have multiplied.

Ⓒ Self-propelled wheelchairs are now common.

Ⓓ The number of injuries involving stairs has increased.

© Harcourt

12. The table below lists properties of some elements in the periodic table.

Element	State of Matter at Room Temperature	Atomic Number
Aluminum	solid	13
Argon	gas	18
Boron	solid	5
Bromine	liquid	35
Cobalt	solid	27
Fluorine	gas	9
Iron	solid	26
Lithium	solid	3

Which of the elements listed are solid at room temperature **and** have atomic numbers less than 10?

Ⓐ Boron and Lithium

Ⓑ Boron and Fluorine

Ⓒ Aluminum and Boron

Ⓓ Aluminum and Argon

13. A team of students will be building a car. The car must carry one team member as it goes around a track during a competition. The team may spend only a certain amount of money. What can the team do to help make sure the final car works, without spending too much money?

   (A) Build cars until one works.

   (B) Build cars until the money runs out.

   (C) Design and build a working model first.

   (D) Use only the lightest, cheapest materials.

14. Which of the following causes earthquakes and volcanic eruptions?

   (A) erosion and deposition

   (B) meteorological events

   (C) plate movement

   (D) vibrations in Earth's core

© Harcourt

Name _____

Date _____

15. A septic system is a sewer system for one building. The diagram shows the basic plan of a septic system. Inside the tank, bacteria break down wastes. Water with broken-down wastes goes through pipes into the drain field.

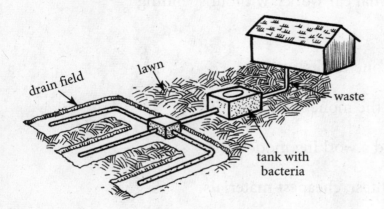

Some houses with septic systems have a patch of grass that is much thicker and greener than the grass in any other part of the yard. Where is this patch of grass, and why is it there?

(A) over the septic tank because of the raw wastes

(B) far away from the septic system because it is cleaner there

(C) right next to the house because the temperature is the warmest there

(D) over the drain field because of nutrients from wastes the bacteria broke down

© Harcourt

16. Veronica made the doorbell shown below using an
    electromagnet.

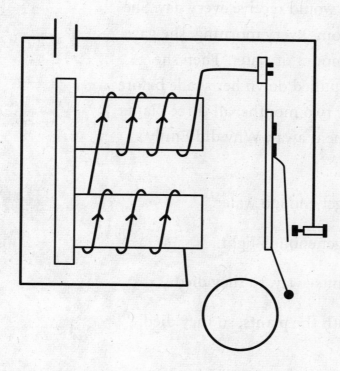

Why did she use an electromagnet rather than a
permanent magnet?

_____

_____

_____

After making the doorbell, she found that it didn't ring
very loudly. How could she change the design of her
doorbell to send more energy to the bell?

_____

_____

© Harcourt

17. Emily wanted to test how much water a certain type of plant needs. She labeled three similar plants according to how much water each would receive every day. She kept the plants in her room. Every morning, she gave each plant the correct amount of water. Then she turned her light off and pulled down her shade before she went to school. After two months, all three plants died, and Emily threw them away. Why did Emily's plants die?

Ⓐ The plants did not get enough water.

Ⓑ The plants did not get enough light.

Ⓒ Emily threw the plants away, so they died.

Ⓓ Emily was bored with the plants, so they died.

18. Look at the illustration of a flower.

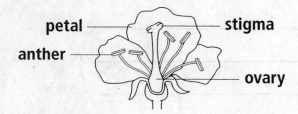

On which part of the flower is pollen made?

Ⓐ anther   Ⓒ petal

Ⓑ ovary   Ⓓ stigma

© Harcourt

19. Louis and Floyd are playing on a seesaw. Louis gets on first. When Floyd climbs on, the seesaw raises Louis up. What force made the seesaw raise Louis up?

(A) electricity

(C) gravity

(B) friction

(D) thrust

20. A substance with a lower density floats on a substance with a higher density. Look at the table below.

Densities of Common Liquids	
Liquid	Density (g/ml)
Cocoa butter	0.96
Coconut Oil	0.24
Glycerin	1.26
Isopropyl alcohol	0.78
Olive oil	0.92
Pure water	1.00
Sea water	1.03

If you pour all of these liquids into a jar and wait a few minutes, which liquid will be in a layer beneath the others?

(A) Coconut oil

(C) Olive oil

(B) Glycerine

(D) Pure water

21. A population of rabbits lives in a thicket of blueberry bushes. They eat leaves from the bushes. A rabbit can eat leaves as high as it can reach while standing on its hind legs. There are few predators, so the rabbit population increases. All the rabbits compete for the same food source (blueberry leaves).

Explain which rabbits are most likely to survive and pass their traits on to their offspring?

_____

_____

_____

Explain how the traits of the rabbits will likely change over time if the environment doesn't change.

_____

_____

_____

22. Charlie is trying to figure out why his guppies keep dying. He decides to test how different types of water affect the growth of the fish. Charlie's setup at the beginning of his experiment is shown below. Each tank has one fish. He feeds the fish and records their lengths each day. If the water in a tank gets low, he adds the correct type of water.

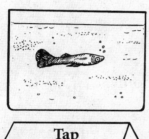 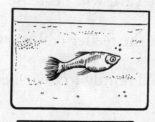 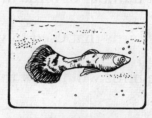

Tap **Distilled** **Bottled**

What should Charlie have done differently in his experiment?

Ⓐ Keep all fish in the same tank of water.

Ⓑ Use three different types of fish.

Ⓒ Record the size of the fish every week.

Ⓓ Start with fish that were about the same size.

23. A family keeps a bird feeder. Many kinds of seed-eating birds use the feeder. Family members have counted as many as 80 birds in their yard at one time. The family moves away, and the next family to live in the house does not keep a bird feeder. What will happen?

　Ⓐ　The birds will begin to eat grass instead of seeds.

　Ⓑ　The number of birds in the yard will decrease.

　Ⓒ　The birds that used to visit the feeder will starve.

　Ⓓ　The birds will follow the old family to their new home.

24. *Amps* are used to describe which of the following?

　Ⓐ　amount of electrical energy

　Ⓑ　amount of electrical pressure

　Ⓒ　electrical power in a system

　Ⓓ　rate of electrical current flow

25. The diagram shows a cross-section of an underground cavern.
The rock surrounding the cavern is limestone.

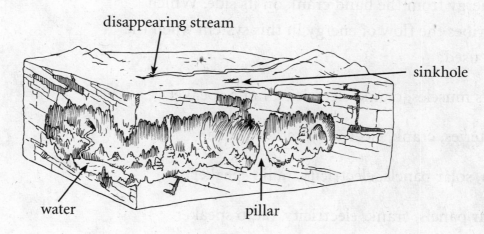

disappearing stream

sinkhole

water

pillar

How did the cavern form?

_____

_____

_____

How did the sinkhole form?

_____

_____

_____

26. Joe has a radio that runs two different ways. It can run off energy captured by its solar panels, or it can run using energy from the hand crank on its side. Which list describes the flow of energy in this system when the crank is used?

    Ⓐ Joe's muscles, crank, electricity, radio speaker

    Ⓑ batteries, crank, electricity, radio speaker

    Ⓒ sun, solar panels, electricity, radio speaker

    Ⓓ solar panels, crank, electricity, radio speaker

27. The symbol used in the periodic table for the element *iron* is derived from the element's Latin name. Which of the periodic table boxes shown below describes *iron*?

    Ⓐ
    14
    Si
    Iron

    Ⓒ
    18
    Ar
    Iron

    Ⓑ
    3
    Li
    Iron

    Ⓓ
    26
    Fe
    Iron

© Harcourt

28. Some students in a class use a balance to compare the masses of different objects. The table shows the number of plastic blocks needed to balance each object.

Blocks Needed to Balance Objects	
Object	Number of Blocks
CD case	8
Calculator	7
Eraser	6
Stapler	9

Each plastic block has a mass of 6.4 g. Which object has a mass of about 45 grams?

(A) CD case

(B) calculator

(C) eraser

(D) stapler

© Harcourt

29. Button batteries are common in watches and other small electric devices. Many button batteries contain the element mercury, which is poisonous. When batteries go into the trash, mercury goes into the trash, too. If the trash is burned, mercury gets into the air, and then it falls in rain. If the trash is buried, mercury gets into the soil.

    Which method for reducing mercury in the environment would probably work **best**?

    Ⓐ Completely ban the use of button batteries.

    Ⓑ Completely ban the burning or burying of trash.

    Ⓒ Keep button batteries out of the trash through recycling.

    Ⓓ Stop selling watches and other small electric devices.

30. A blob of liquid rock, or lava, shoots out of a volcano and lands in the ocean. Almost instantly, the lava hardens into a blob-shaped solid. Which property is the same before and after the lava blob hardens into a solid?

    Ⓐ density

    Ⓑ flexibility

    Ⓒ mass

    Ⓓ temperature

© Harcourt

31. The table lists properties of different objects.

Object	Mass (g)	Volume (cm³)
1	25.4	7.25
2	20.1	8.03
3	32.2	7.16
4	25.3	7.09

Which of the following arranges the objects according to their volume?

Ⓐ 2, 1, 4, 3

Ⓑ 4, 3, 1, 2

Ⓒ 3, 2, 1, 4

Ⓓ 2, 4, 1, 3

32. The illustration below shows how an earthquake occurs
when Earth's crust moves and releases energy. The *focus*
is the point at which the earthquake originates.

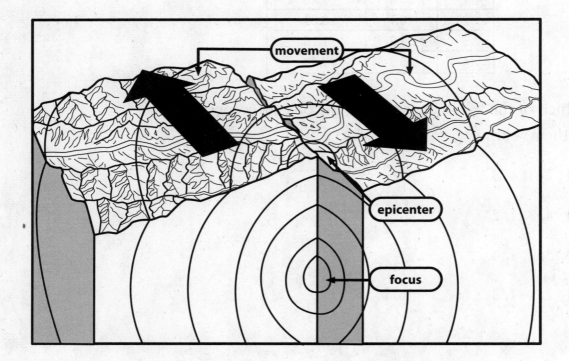

In which of Earth's layers is the focus located?

(A) crust

(B) mantle

(C) outer core

(D) inner core

© Harcourt

Read each clue. Use the terms in the box to complete the puzzle.

ACROSS

4. A living thing, such as a plant, that makes its own food

5. An animal that eats other animals; sometimes called a second-level consumer

7. The process in which plants make food using water from the soil, carbon dioxide from the air, and energy from sunlight

8. A consumer that obtains food energy by breaking down the remains of dead plants and animals

DOWN

1. Diagram that shows how much food energy is passed to each level in a food chain

2. A community of organisms and the environment in which they live

3. An animal that eats producers

6. An animal that eats plants or other animals

carnivore
consumer
decomposer
ecosystem
energy pyramid
herbivore
photosynthesis
producer

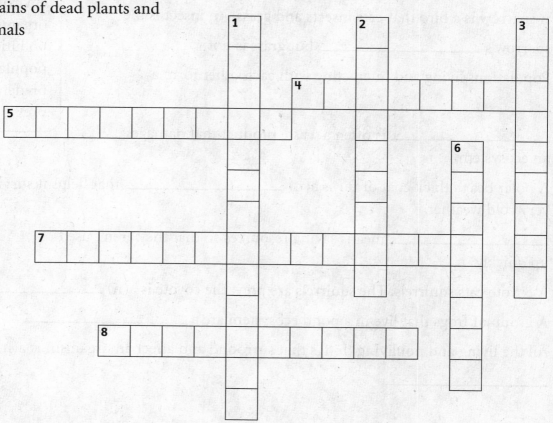

© Harcourt

Name _____ Date _____

Use the terms in the box to label the pictures.

1.

_____

2.

_____

Use the terms in the box to complete the sentences.

1. A(n) _____ is any living thing, such as a tree, an apple, a cat, or a mushroom.

2. A sparrow is a bird that eats insects and grass. An insect is the sparrow's _____, but grass is not.

3. Populations living and interacting with each other form a _____.

4. _____ is often a waste product that damages an ecosystem.

5. A polar bear's thick coat of fur is a(n) _____ that helps it survive in very cold weather.

6. _____ means saving resources so that they aren't used up quickly.

7. A coyote eats squirrels. The squirrels are prey; the coyote is a(n) _____.

8. A group of frogs that live in a pond ecosystem are a _____.

9. All the living and nonliving things that surround and affect an organism is a(n) _____.

© Harcourt

Life Science

Match each term in Column B with its meaning in Column A.

Column A

_____ 1. Something so small that it can only be seen by using a microscope

_____ 2. A single-celled organism with a nucleus and organelles

_____ 3. A group of tissues that work together to perform a certain function

_____ 4. Cells that work together to perform a certain function

_____ 5. Stages that living things pass through as they grow and change

_____ 6. An area where an organism can find everything it needs to survive

Column B

A. protist

B. life cycle

C. microscopic

D. tissue

E. organ

F. habitat

Use the terms in the box to label the diagram.

consumer
decomposer
producer

7. _____

8. _____   9. _____

Name _____ Date _____

Read each clue. Use the terms in the box to complete the sentences.

1. _____ forms when melted rock cools and hardens.

2. _____ forms when layers of rock are squeezed and become stuck together.

3. _____ forms when high heat and great pressure change existing rock.

> igneous rock
> metamorphic rock
> sedimentary rock

Write each term from the box in the correct column of the chart.

Part of Our Solar System	Not Part of Our Solar System

> constellation
> galaxy
> moon
> planet
> sun

Read the clues. Use the terms in the box to complete the sentences.

1. I have a star at my center, and planets revolve around it.

   I am a(n) _____.

2. The moon looks different each night because of me.

   I am a(n) _____.

3. I'm an imaginary line that passes through Earth's poles.

   I am Earth's _____.

4. I'm an imaginary line equally distant from both the North and South Poles.

   I am Earth's _____.

5. I'm a huge ball of very hot gas. I am a(n) _____.

6. I'm everything that exists, including stars, planets, and dust.

   I am the _____.

7. I am the bending of light. I am _____.

> axis
> equator
> moon phase
> refraction
> solar system
> star
> universe

Name _____ Date _____

Read each clue. Use the terms in the box to complete the puzzle.

atmosphere
condensation
deposition
earthquake
erosion
evaporation
topography
volcano
water cycle
weathering

ACROSS

3. All the kinds of landforms in a certain place

5. The process of moving sediment by wind, moving water, or ice

8. The process by which a gas changes into a liquid

9. The process of wearing away rocks by natural processes

10. The blanket of air surrounding Earth

DOWN

1. The process in which sediment settles out of water or is dropped by the wind

2. The continuous movement of water from Earth's surface into the atmosphere and back again

4. A mountain made of lava, ash, or other materials from eruptions that occur at an opening in Earth's crust

6. The process of a liquid changing into a gas

7. Movement of the ground caused by a sudden release of energy in Earth's crust

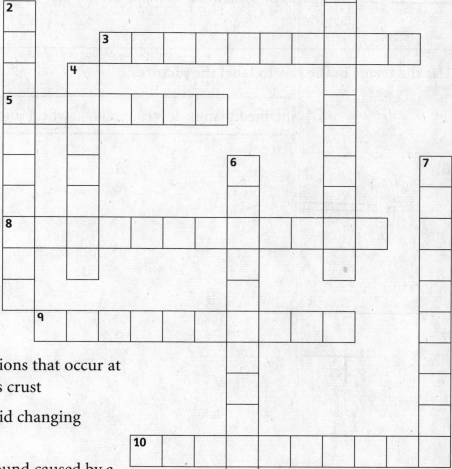

© Harcourt

Name _____ Date _____

Match each term in Column B with its meaning in Column A.

Column A

_____ **1.** A resource that, once used, cannot be replaced in a reasonable amount of time

_____ **2.** A resource that can be replaced within a reasonable amount of time

_____ **3.** The remains or traces of past life found in sedimentary rock

_____ **4.** Anything that dirties or harms the environment

_____ **5.** Using less of something to make the supply last longer

Column B

A. conservation

B. fossil

C. nonrenewable resource

D. pollution

E. renewable resource

Use the terms in the box to label the pictures.

> inclined plane   lever   pulley   wheel-and-axle

6.
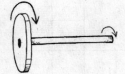

_____

8.
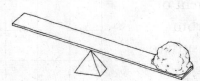

_____

7.
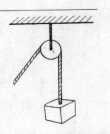

_____

9.
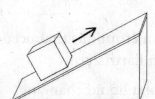

_____

Name _____ Date _____

Read each clue. Use the terms in the box to complete the puzzle.

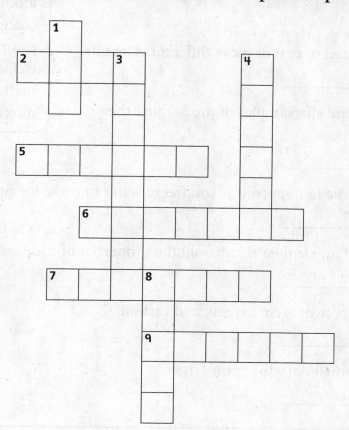

density
gas
liquid
mass
matter
mixture
solid
solution
volume

ACROSS

2. The amount of matter in any object is its _____.

5. The _____ of an object is the amount of space it takes up.

6. You create a(n) _____ if you pour oil into a glass of water.

7. The concentration of matter in an object is its _____.

9. If you drink a glass of water, the water is in the _____ state.

DOWN

1. Matter without a definite shape or volume is a(n) _____.

3. You create a(n) _____ if you pour drink powder into a glass of water and the water turns red and is cherry-flavored.

4. A glass of milk, a rock, and air are all examples of _____ because they take up space and have mass.

8. Matter that has a definite shape and a definite volume is a(n) _____.

© Harcourt

Name _____   Date _____

Read the clues, then use the terms in the box to complete the sentences.

atom
change of state
compound
element
matter
microscope

1. I am a substance composed of two or more different elements.

   I am a(n) _____.

2. Water, oxygen, and dirt are all examples of me because they have mass and take up space.

   I am _____.

3. One way you can cause me to happen is if you freeze water to make ice cubes.

   I am a(n) _____.

4. I am the smallest unit of an element that has all the properties of the element.

   I am a(n) _____.

5. Some examples of me are hydrogen, oxygen, and carbon.

   I am a(n) _____.

6. I am a tool that makes small objects appear larger.

   I am a(n) _____.

Use the terms from the box to label the pictures.

chemical change
physical change

7.

   _____

8.

   _____

© Harcourt

Use the terms in the box to complete the sentences.

| energy |
| energy transfer |
| experiment |
| investigation |
| kinetic energy |
| potential energy |
| reflection |
| scientific method |
| system |

9. When energy moves from one place or object to another, a(n) _____ takes place.

10. A soccer ball rolling across a field has _____, or energy of motion.

11. Two ways that you can increase the _____ of a ball are to throw it or to raise it above your head.

12. Moving a book to a higher shelf increases the book's _____ because you have changed the book's position.

13. A(n) _____ is a procedure you carry out under controlled conditions to test a hypothesis.

14. A(n) _____ is a group of separate elements that work together to accomplish something.

15. The bouncing of heat or light off an object is called _____.

16. A procedure that is carried out to gather data about an object or event is called a(n) _____.

17. The _____ is a series of steps that scientists use when performing an experiment.

Read each clue. Use the terms in the box to complete the puzzle.

force
friction
fulcrum
gravitational force
gravity
heat
inertia
light
magnetism

ACROSS

3. The force of attraction between Earth and objects on or near Earth

5. The force produced by a magnet

7. Radiation from the sun that we see

8. A force that opposes motion

9. The property of matter that keeps it moving in a straight line or keeps it at rest

DOWN

1. The pull of all objects in the universe on one another

2. The transfer of thermal energy between objects with different temperatures

4. The balance point on a lever that supports the arm but does not move

6. A push or pull that causes an object to move, stop, or change direction

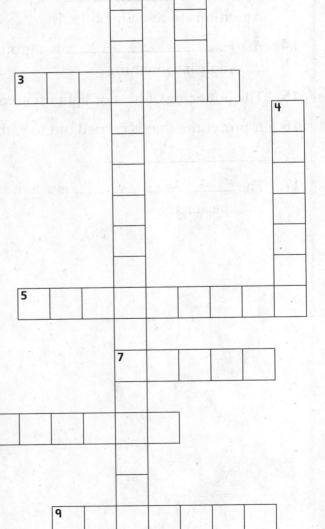

© Harcourt